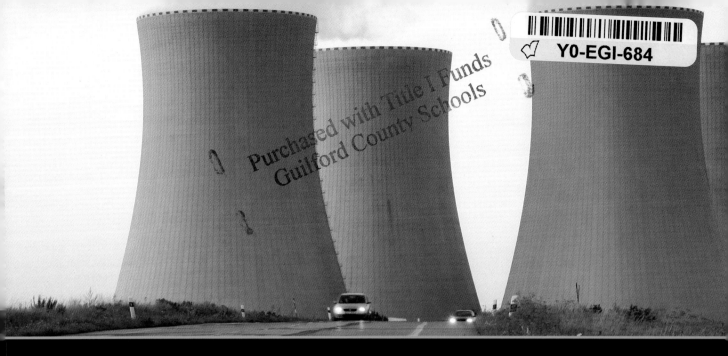

Energy Resources

by Laura McDonald

What will we use to power the world?

TABLE OF CONTENTS

4 INTRODUCTION
The Biggest Dam in the World

China proposed a huge dam to control flooding and generate electricity. People in and outside of China worried about the damage the planned dam could cause. Which would prevail—the quest for energy or the needs of local people and wildlife?

6 CHAPTER 1
What Are Energy Resources?

How do people choose among and use energy resources?

16 CHAPTER 2
Energy from Fossil Fuels

What are the advantages and disadvantages of energy from fossil fuels?

Cartoonist's Notebook..**24**

26 CHAPTER 3
Nuclear Energy

What are the advantages and disadvantages of nuclear energy?

32 CHAPTER 4
Renewable Energy Resources

What are the advantages and disadvantages of renewable energy resources?

42 CONCLUSION
Quest for Energy..**42**

How to Write a Persuasive Letter..........................**44**

Glossary..**46**

Index..**48**

What are the advantages and limitations of our planet's energy resources?

INTRODUCTION

The Biggest Dam

CHINA HAS 1.3 BILLION PEOPLE. All those people need energy to run factories, drive cars, and power homes. Where will China find enough energy?

China has used coal as a major energy resource. Coal mining is dangerous. Burning coal makes a lot of pollution. China also imports petroleum, or crude oil, from other countries. China has many rivers. The energy of moving river water is clean.

The Yangtze (YANG-see) is one of those rivers. The river carries water from the mountains to the city of Shanghai. The Yangtze also brings floods. The Yangtze flooded 1,000 times in the last 2,000 years. In 1931, a flood killed 140,000 people.

The Chinese would like to control the Yangtze River. Their biggest plan to control the river was the Three Gorges Project. This plan would build the largest dam in the world. Water would pour through the dam. This moving water would make a lot of electricity.

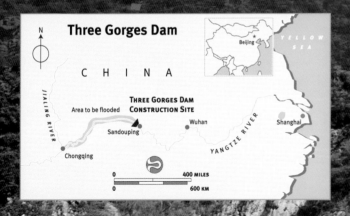

▲ The Three Gorges Project would flood a large area while producing electricity and controlling the Yangtze River.

in the World

Many people did not like the Three Gorges Project. Millions of people live near the Yangtze. Water trapped by the dam would flood many towns. Where would people go?

Many plants and animals also live along the Yangtze. The Chinese river dolphin lives only in the Yangtze River. People did not think that the dolphin could survive the dam.

China had a big challenge. Leaders had to find a way to make clean energy and control floods. But China also had to take care of local people and wildlife.

Coal-burning power plants cause much of the smog in China.

CHAPTER 1
What Are Energy Resources?

Energy powers our world. Cars, televisions, toasters, and factories need energy to work. What is energy? What sources of energy can people use?

▲ People use energy for many different purposes.

What Is Energy?

Energy (EH-ner-jee) is the ability to do work. Light and heat are forms, or types, of energy. Sound and electricity are also forms of energy. Look at the picture below. Energy moves the cars down the street. Lamps make light energy. Even trees use energy to grow.

Stored energy is called **potential energy** (puh-TEN-shul EH-ner-jee). Food, wood, and gas all have potential energy. **Kinetic energy** (kih-NEH-tik EH-ner-jee) is the energy of movement. Flowing water has kinetic energy. Blowing air has kinetic energy, too. Think of a cyclist riding down a hill. As she goes faster, she gains kinetic energy.

Energy can change from one form to another. Each time energy changes form, some of it turns into heat.

Essential Vocabulary

- energy — page 7
- energy resource — page 8
- kinetic energy — page 7
- nonrenewable resource — page 8
- potential energy — page 7
- renewable resource — page 8

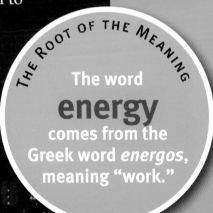

THE ROOT OF THE MEANING

The word **energy** comes from the Greek word *energos*, meaning "work."

How do people choose among and use energy resources?

CHAPTER 1

Energy Sources

Energy comes from many different sources. An energy source that people use is called an **energy resource** (EH-ner-jee REE-sors).

The sun is our most important source of energy. Sunlight, or solar energy, is a **renewable resource** (rih-NOO-uh-bul REE-sors). Nature can replace it quickly. The sun also helps make the energy of wind, waves, and running water.

Plants turn the kinetic energy of light into the potential energy of food. The food gives living things energy to grow and reproduce. Animals gain this energy when they eat plants. People use plant and animal materials, or biomass, as energy resources, too.

We also use the energy from things that lived long ago. Coal, natural gas, and petroleum are all fuels. These fuels come from the remains of living things. We call them fossil fuels. These fuels are **nonrenewable resources** (nahn-rih-NOO-uh-bul REE-sors-ez). These resources cannot be replaced.

The heat inside Earth is an energy resource that does not come from the sun. Another nonsolar energy resource is nuclear energy. Nuclear energy is the energy stored in atoms.

WHAT ARE ENERGY RESOURCES?

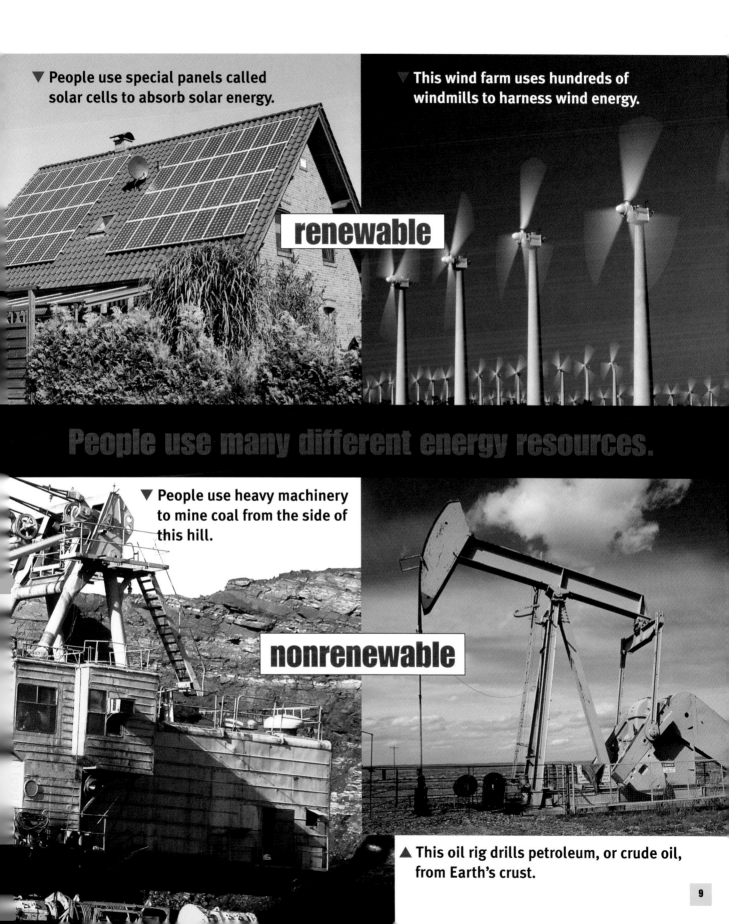

▼ People use special panels called solar cells to absorb solar energy.

▼ This wind farm uses hundreds of windmills to harness wind energy.

renewable

People use many different energy resources.

▼ People use heavy machinery to mine coal from the side of this hill.

nonrenewable

▲ This oil rig drills petroleum, or crude oil, from Earth's crust.

CHAPTER 1

Using Energy Resources

Energy resources are all around us. It takes some work to use the resources. First, people must get the energy from resources. People build dams to collect moving water. People use pumps to get oil from the ground. People use solar cells to store sunlight.

Next, pipelines, trucks, ships, and trains must move energy resources to the places where they will be used. Then people process energy resources to make useful fuels. Petroleum, or crude oil, starts out as a thick, gooey liquid. We separate the thick petroleum into gasoline, heating oil, and other fuels. We make nuclear fuel from uranium (yuh-RAY-nee-um) that people find underground. We can process biomass and fossil fuels into clean hydrogen gas.

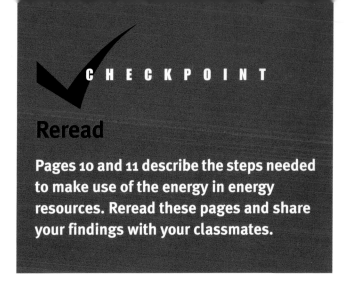

✓ CHECKPOINT

Reread

Pages 10 and 11 describe the steps needed to make use of the energy in energy resources. Reread these pages and share your findings with your classmates.

Moving water is stored behind a dam so that its energy can be used.

Converting Energy

People use machines to convert, or change, resources into useful energy. Car engines burn gas. The gas has potential energy. This becomes the kinetic energy of a moving car.

Power plants convert energy resources to make electricity. Moving water or burning fuel turns a giant fan called a turbine. The turbine spins a generator. The generator makes electricity. The electricity can be stored in batteries. It can also be sent through power lines. The power lines bring the electricity to homes. They also bring it to businesses.

Each step to change and move energy also uses energy. The energy that goes into these steps is lost as heat.

FROM ENERGY RESOURCE TO USEFUL WORK

COLLECTION

TRANSPORTATION

PROCESSING

TRANSPORTATION

CONVERSION TO USEFUL WORK

CHAPTER 1

Choosing Among Energy Resources

How do people choose which energy resources to use? People often use the cheapest energy resource they can find.

Long ago, people in Europe used wood for heat. Wood soon became more and more scarce. The lack of wood drove up the price. By the 1800s, the price of wood was higher than the price of coal. So coal replaced wood. Today we are interested in "alternative" fuels. This is because gasoline and diesel are getting more expensive.

The price of an energy resource is not the only factor. Energy choices also affect safety, people's health and happiness, and the environment. Many people like energy resources found in their own country. People also want clean energy resources that won't pollute the air and water. Pollution harms all living things.

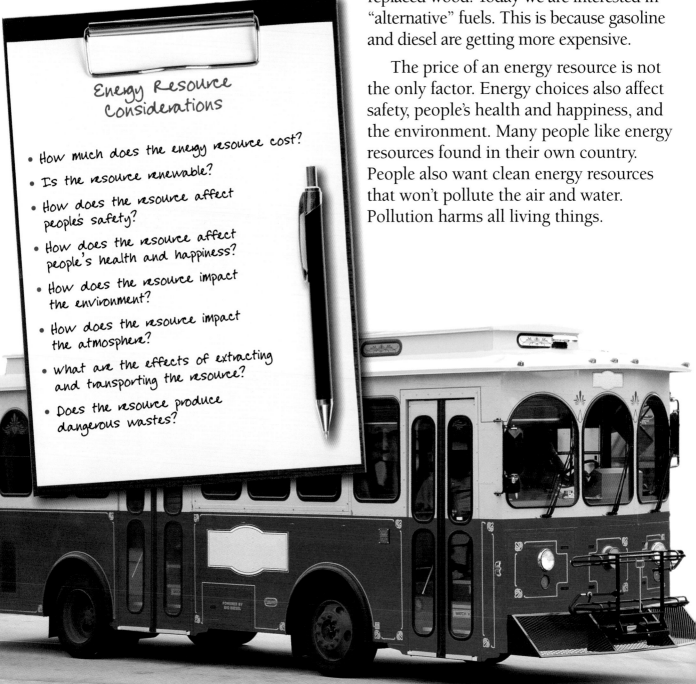

Energy Resource Considerations

- How much does the energy resource cost?
- Is the resource renewable?
- How does the resource affect people's safety?
- How does the resource affect people's health and happiness?
- How does the resource impact the environment?
- How does the resource impact the atmosphere?
- What are the effects of extracting and transporting the resource?
- Does the resource produce dangerous wastes?

▲ When the costs of one energy resource increase, people start to explore new resources.

WHAT ARE ENERGY RESOURCES?

Burning fossil fuels adds too much carbon dioxide gas to the air. This is a form of pollution. Too much of this gas in the air adds to global climate change. The air and water will get warmer. All living things will have to adjust to new weather patterns.

Scientists want to slow global climate change. Switching to renewable energy resources might help.

We can also reduce carbon dioxide by being more efficient. Efficiency is how much useful work you get from an amount of energy. If a car can go twice as far on a tank of gas, it is twice as efficient. The car also makes half as much carbon dioxide.

✓ CHECKPOINT

Talk It Over

The cost of using gasoline is much higher than the price people pay at the pump. Hold a class debate about whether the price paid for energy resources should include all the costs, from extraction to disposal of wastes.

You need to know all the effects of using energy resources to use them well. The next three chapters tell you about the good and bad qualities of each of our energy resources.

▼ Polar bears depend on sea ice, which is shrinking due to global climate change.

CHAPTER 1

Hands-On Science

Feel the Heat

Do you need a new lightbulb? In the store, you can choose between regular incandescent lights and compact fluorescent lights (CFLs). Which type of lightbulb uses electricity more efficiently?

TIME REQUIRED
20 minutes

MATERIALS NEEDED
- desk or table lamp
- compact fluorescent light (CFL) and incandescent lightbulb with the same light output/wattage
- thermometer
- ring stand
- centimeter ruler

SAFETY CONSIDERATIONS

Incandescent lightbulbs get very hot—do not touch the bulb after the experiment. Turn off the lamp before screwing in or removing a lightbulb.

PROCEDURE

1. Make sure the lamp is turned off. Screw in the CFL bulb. Turn on the lamp and observe the light produced.

2. Use a ring stand to hold the thermometer 15 cm (6 in) away from the glowing light. Record the temperature on the thermometer. After 10 minutes, record the temperature again.

3. Turn off the lamp and allow the bulb and thermometer to cool for 3 minutes. Remove the bulb.

4. Repeat steps 1 and 2 with the incandescent bulb. Turn off the lamp.

ANALYSIS

1. Compare the heat and light produced by each lightbulb.

2. Which lightbulb is more efficient at converting electricity to light?

3. Is it possible to create a lightbulb that produces only light and no heat? Explain your answer.

DATA TABLE

LIGHT TYPE	DESCRIBE THE LIGHT PRODUCED	TEMPERATURE MEASUREMENTS	CHANGE IN TEMPERATURE
Compact fluorescent light (CFL)		Start: Finish:	
Incandescent light		Start: Finish:	

WHAT ARE ENERGY RESOURCES?

SUMMING UP

- Energy resources provide people with the energy they need for modern life.
- People must collect, extract, transport, and process most energy resources before their energy can be used.
- Each energy resource offers advantages but also comes with limitations.
- People and governments consider many factors before they choose which energy resources to use.

Putting It All Together

Choose one of the activities below.

1. A magazine article claims to have found a perfect, free energy resource. Write a letter to the editor of the magazine responding to this claim.

2. Walk around your home and make a list of all the appliances and machines that require energy to work. Include the energy source for each item on your list. Compare your list with a friend's list.

3. Reread pages 12 and 13, and then predict what would happen if the price of petroleum became extremely high. Write a paragraph explaining how your life might be different after people adjusted to the high cost of petroleum.

4. You read on page 8 that sunlight causes the movement of wind. Research the energy conversions that lead to the creation of wind. Make a diagram and share your findings with your class.

CHAPTER 2
Energy from Fossil Fuels

HAVE YOU EVER SEEN A LUMP OF COAL? HAVE YOU EVER SEEN CRUDE OIL, OR A CONTAINER OF NATURAL GAS? AMERICANS USE FOSSIL FUELS FOR 85% OF THEIR ENERGY.

A **fossil fuel** (FAH-sul FYOOL) is an energy resource that is formed inside Earth from ancient plant or animal remains. The biomass of the remains changed over millions of years into coal, natural gas, or petroleum. The diagrams on page 18 show the formation of each type of fossil fuel. Fossil fuels are nonrenewable resources because they take such a long time to form.

Essential Vocabulary

- fossil fuel   page 16

What are the advantages and disadvantages of energy from fossil fuels?

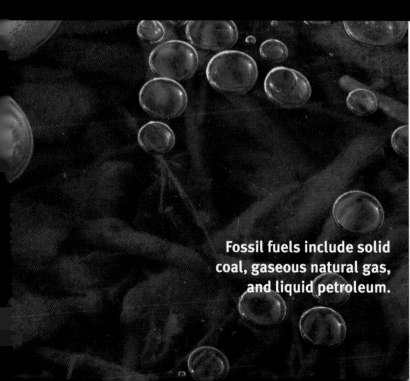

Fossil fuels include solid coal, gaseous natural gas, and liquid petroleum.

CHAPTER 2

Fossil fuels have a lot of potential (stored) energy. Burning fossil fuels turns this energy to heat energy. We can use this energy to warm buildings. We can use it to make electricity or to power factories. Fossil fuels are very helpful. But fossil fuels can also be very harmful.

✓ CHECKPOINT

Read More About It

Read more about the ancient forests and swamps that became the coal deposits of today. How were these forests like the forests of today? How were they different?

How Coal Formed

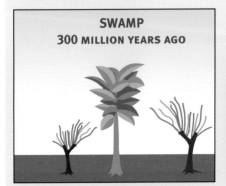

Before the dinosaurs, many giant plants died in swamps.

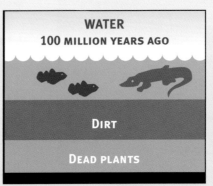

Over millions of years, the plants were buried under water and dirt.

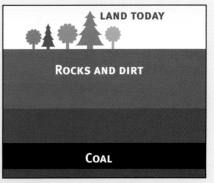

Heat and pressure turned the dead plants into coal.

How Petroleum and Natural Gas Formed

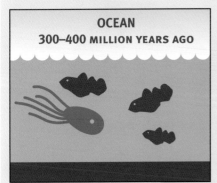

Tiny sea plants and animals died and were buried on the ocean floor. Over time, they were covered by layers of silt and sand.

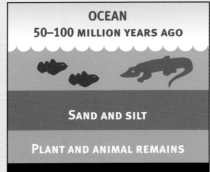

Over millions of years, the remains were buried deeper and deeper. The enormous heat and pressure turned them into oil and gas.

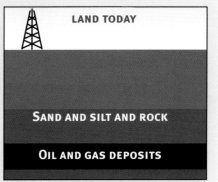

Today, we drill down through layers of sand, silt, and rock to reach the rock formations that contain oil and gas deposits.

ENERGY FROM FOSSIL FUELS

Coal

Coal has many advantages. Coal is easy to move and burn. Coal costs less than most other energy resources. In the United States, coal is common. The country has enough coal to last a hundred years. Today, 50% of electricity in the country comes from coal.

■ U.S. Coal-Producing Regions

Coal also has major disadvantages. Mining accidents and pollution from coal cause thousands of deaths each year. Burning coal is the worst source of air pollution in the world. Burning coal also adds to global climate change. People are trying to find cleaner ways to burn coal.

How Coal Is Used

Industry production (especially steel) 7%

Electricity production 93%

▼ Mining brings coal up to the surface where it can be used.

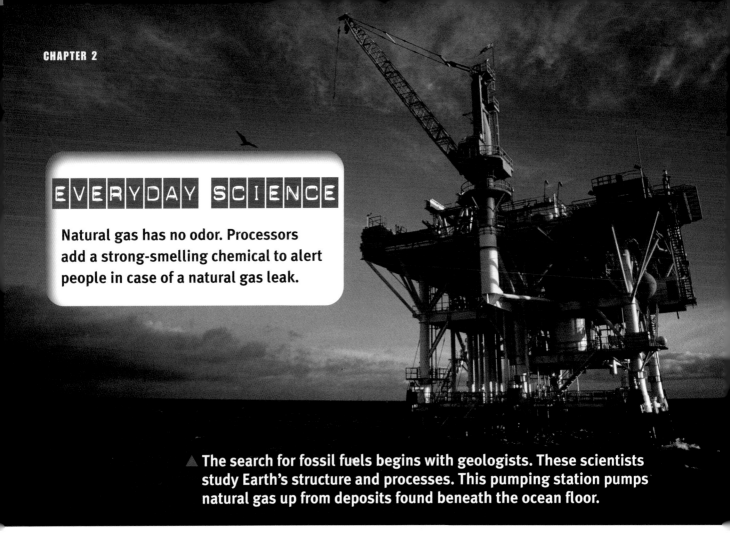

EVERYDAY SCIENCE

Natural gas has no odor. Processors add a strong-smelling chemical to alert people in case of a natural gas leak.

▲ The search for fossil fuels begins with geologists. These scientists study Earth's structure and processes. This pumping station pumps natural gas up from deposits found beneath the ocean floor.

Natural Gas

Natural gas is another cheap energy resource. Burning natural gas turns its potential chemical energy into heat. Natural gas heats buildings, powers factories, and makes electricity. Some new vehicles run on natural gas.

Lots of natural gas is under the ground and in coastal waters. Natural gas even bubbles out of swamps. Natural gas is made mostly of the colorless gas methane. Methane burns easily and cleanly. Methane does not make smoke or ash.

How Natural Gas Is Used

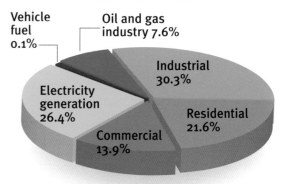

- Vehicle fuel 0.1%
- Oil and gas industry 7.6%
- Industrial 30.3%
- Residential 21.6%
- Commercial 13.9%
- Electricity generation 26.4%

Natural gas use also makes some problems. Burning natural gas makes carbon dioxide. Some natural gas facilities put carbon dioxide back underground. Natural gas explosions have killed many people. But natural gas is much cleaner and safer than coal.

ENERGY FROM FOSSIL FUELS

Petroleum

People around the world use more petroleum (puh-TROH-lee-um) than any other energy resource. Petroleum, or crude oil, is a thick, smelly, dark liquid. Pumps bring petroleum up from the ground or seafloor. Then ships and pipelines move the crude oil to a refinery.

Crude oil has thousands of different chemicals. This mixture is refined, or separated, to make gasoline, plastic, tar, asphalt, food additives, and many other products. Power plants also burn petroleum to make electricity. Petroleum is very useful.

The United States does not have enough petroleum to meet the country's needs. Texas, Alaska, California, and Louisiana produce the most petroleum. The country gets the rest of the crude oil it needs from other countries. Most of these countries are in the Middle East and South America.

Experts disagree about the total amount of petroleum left in the world. This is one reason why the price of petroleum goes up and down so much. We do know that the world's reserves of petroleum are getting smaller every year.

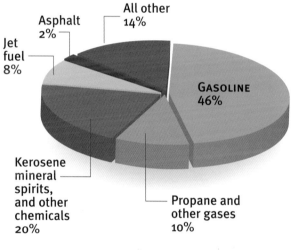

Products Made from Petroleum
- Gasoline 46%
- Kerosene mineral spirits, and other chemicals 20%
- All other 14%
- Propane and other gases 10%
- Jet fuel 8%
- Asphalt 2%

SCIENCE + MATH

Inverse Relationships

An increase in one variable can cause a decrease in another. For example, the more petroleum we take out of the ground, the less there is left. Mathematicians call this an inverse relationship. There is also an inverse relationship between the cost of gasoline and how much people drive. Can you think of another inverse relationship?

Oil spills can cause environmental disasters.

Petroleum also has its disadvantages. Drilling for oil could seriously hurt the local wildlife. Burning petroleum makes air pollution. Ships and pipelines that carry crude oil can leak or spill oil into the ocean. In the worst accidents, millions of liters of petroleum coat the water with a thick layer of oily gunk. Fish, seabirds, small animals, and plants die in large numbers.

Many crude oil deposits are in unstable countries. This means protecting the oil supply is both costly and dangerous.

The problems with petroleum are forcing the world to look for other ways. People want to find a mix of energy resources to replace petroleum. This is a big challenge.

Advantages and Disadvantages of Fossil Fuels

ENERGY RESOURCE	ADVANTAGES	DISADVANTAGES
Coal	• Large supply available in the United States • One of the least expensive energy resources • Easily transported by trains	• Nonrenewable resource • Mining damages environment • Mining accidents result in many deaths • Can produce air and water pollution
Natural Gas	• Clean fuel • Requires little processing • Large supply available in the United States	• Nonrenewable resource • Contributes to global climate change by creating carbon dioxide • Highly flammable • Costly to transport
Petroleum	• Contains many useful fuels • Easily transported liquid • Relatively inexpensive to produce	• Nonrenewable resource • Worldwide supply shrinking rapidly • Extraction, transportation, and burning cause environmental damage • Contributes to global climate change • Many deposits in politically unstable countries

ENERGY FROM FOSSIL FUELS

SUMMING UP

- Fossil fuels form from the remains of ancient plants and animals.
- Petroleum, coal, and natural gas are extracted from mines and wells underground or under the sea.
- People burn these nonrenewable energy resources to power transportation, generate electricity, and produce heat.
- Modern society relies heavily on nonrenewable resources because they are inexpensive, convenient energy resources.
- However, supplies of fossil fuels are limited and their use can endanger human life and damage the environment.

Putting It All Together

Choose one of the activities below.

1 You hear a caller on a radio show suggest that drilling for oil in Alaska would solve the need for more petroleum. How would you respond to this caller? Write a paragraph explaining your response. Compare your response to a classmate's.

2 The coal industry is trying to reduce air and water pollution by creating clean coal technology. Read more about clean coal technology in a library or on the Internet. Show some of the new developments on a poster to share with your class.

3 Reread page 21 about the uses of petroleum. For one day, record all the times you use or observe items made from or powered by petroleum products.

Believe it: Any car with a diesel engine can be modified into a "grease car" that runs on WVO — Waste Vegetable Oil — thrown out from restaurants. The car will get the same gas mileage and won't add extra carbon dioxide to the atmosphere.

True: Solar cookers use the sun as their sole source of energy. There is no cost to run them, nor to the environment. Many of the dozens of types of solar cookers can be built from everyday materials in just a few hours.

Stinky but True: Cow waste contains methane, a colorless, though alas, not odorless gas. This gas can be converted to electricity.

Well... this one isn't true — YET!

Many alternative energy sources may seem crazy at first, but some of these ideas actually work.

Can you think of other ideas that people thought were wacky at first, but then changed history?

CHAPTER 3
Nuclear Energy

THE CORE OF AN ATOM IS CALLED A NUCLEUS (NOO-KLEE-US). A NUCLEUS HAS PROTONS AND NEUTRONS. THE NUCLEUS CONTAINS MANY POSITIVE CHARGES. A FORCE, CALLED THE STRONG FORCE, HOLDS EVERYTHING IN THE NUCLEUS TOGETHER.

Using Nuclear Energy

Some types of atomic nuclei (NOO-klee-i) are **radioactive** (ray-dee-oh-AK-tiv). This means they can decay, or change into other types of nuclei. As each nucleus decays, it releases energy. This energy is a form of radiation. Some nuclei split into two smaller nuclei and some particles. This is called **fission** (FIH-shun). This is how we get nuclear energy.

Uranium-235 is radioactive. This type of uranium is commonly found in rocks all over the world.

Uranium-235 nuclei split to form the nuclei of two different atoms, as well as two or three neutrons. These neutrons can be taken in by other uranium-235 atoms. This causes them to become unstable. Then they also undergo fission. As more and more neutrons are released, more and more uranium atoms are split. This is called a chain reaction. The sum of all the energy that is released produces a big explosion.

Essential Vocabulary

- **fission** — page 27
- **nuclear power** — page 28
- **nucleus** — page 26
- **radioactive** — page 27

What are the advantages and disadvantages of nuclear energy?

A domino race is an example of a chain reaction.

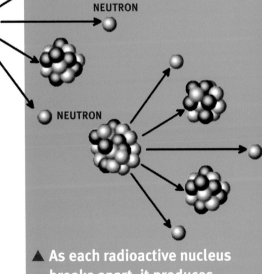

▲ As each radioactive nucleus breaks apart, it produces two or three neutrons that collide with other nuclei.

CHAPTER 3

In a power plant, nuclear energy heats water to produce steam. The kinetic energy of steam can move nuclear submarines and warships. The steam can also spin turbines to make electricity. In the photograph below, you can see steam rising from cooling towers. Nuclear energy makes 15% of the electricity in the world. Electricity made using nuclear energy is called **nuclear power** (NOO-klee-er POW-er).

A building called a reactor holds the radioactive uranium fuel at a nuclear power plant. The white dome in the photograph below is the reactor. Almost no radiation escapes from the reactor.

Nuclear power plants are very expensive to build. Nuclear power plants last for decades and are cheap to run. Nuclear power plants give off little pollution and produce no carbon dioxide. Nuclear energy is a nonrenewable resource, but the world's uranium supply should last for centuries.

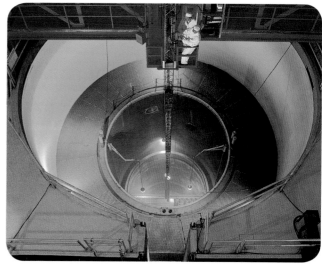

▲ A nuclear reactor emits a beautiful blue glow.

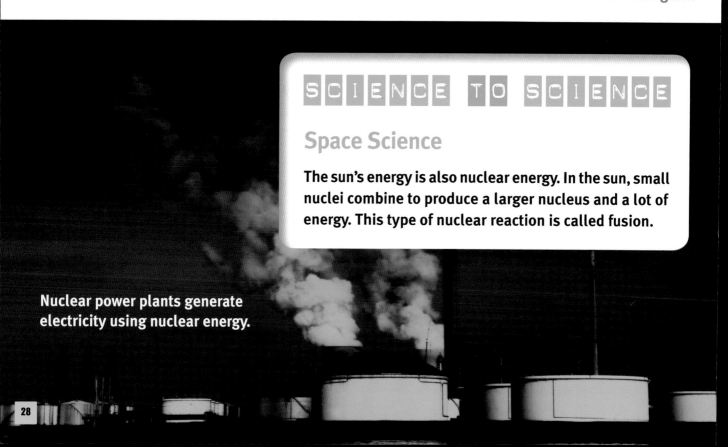

Nuclear power plants generate electricity using nuclear energy.

SCIENCE TO SCIENCE

Space Science

The sun's energy is also nuclear energy. In the sun, small nuclei combine to produce a larger nucleus and a lot of energy. This type of nuclear reaction is called fusion.

NUCLEAR ENERGY

Hands-On Science

Chain Reaction

You can use dominoes to make a chain reaction. How does your chain reaction compare to a nuclear chain reaction?

TIME REQUIRED
20 minutes

MATERIALS NEEDED
- 15 dominoes
- ruler
- stopwatch or watch with second hand

PROCEDURE

1. Stand 15 dominoes on their ends in a straight row. Set the dominoes about 2.5 cm (1 in) apart. Knock over the first domino with your finger. Time how long it takes for all the dominoes to fall. Record your results. Repeat the experiment two more times and calculate the average time.

2. Arrange the dominoes so that each domino will knock over two others. Use the illustration below as an example. Observe what happens when you knock over the first domino. Time and record how long it takes for the whole set of dominoes to fall over. Repeat the experiment two more times and calculate the average time.

3. Repeat step 2, but this time hold a ruler on its end in the middle of the arrangement. Knock over the first domino and observe what happens.

ANALYSIS

1. How is the domino race similar to a nuclear chain reaction? How are they different?

2. Compare the time it took for all the dominoes to fall in step 1 and step 2. Explain any differences you observed.

3. What effect did the ruler in step 3 have on the chain reaction?

DATA TABLE

TIME FOR ALL DOMINOES TO FALL	STEP 1	STEP 2	STEP 3
First trial			
Second trial			
Third trial			
Average time			

Safety and Waste Disposal

Every energy resource has limits. Nuclear energy can release radioactive materials. Radiation can cause burns, cancer, or death.

Accidents

An uncontrolled nuclear reaction can make enough heat to cause a "meltdown." A meltdown destroys the reactor and puts radioactive material into the environment. Nuclear reactors have rods that absorb neutrons and slow down the reaction. Water cools the reactor.

There have been two meltdowns in over fifty years of nuclear power. In 1979 the Three Mile Island plant in Pennsylvania melted down but caused no injuries. A more serious meltdown occurred in 1986 at the Chernobyl (cher-NOH-bul) plant in Ukraine. Dozens of rescue workers died and many local residents developed cancer. The Chernobyl meltdown killed about sixty people.

Nuclear Waste

Used nuclear fuel and other wastes can stay dangerously radioactive for decades or even thousands of years. Now we store nuclear wastes at nuclear power plants. Some nuclear wastes can be turned into new fuels. Still, nuclear power plants continue to make radioactive waste.

Nuclear Terrorism

Governments around the world work to keep radioactive materials secure. People can use nuclear fuel to build a nuclear weapon.

Now, people are planning to build dozens of nuclear power plants in the United States. The rising cost of fossil fuels and concerns about global climate change make many people want to use nuclear energy.

Energy Resource	Advantages	Disadvantages
Nuclear Energy	• Produces no air pollution or carbon dioxide • Large supply of fuel available • Inexpensive to produce after plant is built • Excellent safety record	• Dangerous wastes last for many years • Possibility of accident or terrorist attack • Power plants very expensive to build

SUMMING UP

- Nuclear energy is produced by splitting, or fission, of atoms.
- People use nuclear energy to produce electricity and power ships.
- Nuclear power produces little pollution or carbon dioxide. On the other hand, there are concerns about the safety of nuclear power plants and the handling of their fuel and wastes.
- The United States has yet to decide the role nuclear energy will play in the future.

Putting It All Together

Choose one of the activities below.

1. Research the parts of a nuclear reactor and the purpose of each part. Make a poster to present your findings to your class.

2. Read more about the Yucca Mountain project in a library or on the Internet. Present the arguments for and against this planned nuclear waste storage facility to your class.

3. Lise Meitner (LEE-zuh MITE-ner) was the first scientist to calculate the enormous energy released by splitting an atomic nucleus. Use your library or the Internet to learn more about her interesting life. Discuss her difficult choices with a friend.

✓ CHECKPOINT

Think About It

Even though coal production kills many more people than nuclear power production, people worry more about the dangers of nuclear power. Why do you think that is?

CHAPTER 4
Renewable Energy Resources

FOSSIL FUELS AND NUCLEAR MATERIALS ARE RUNNING OUT. Once people use up these energy resources, they cannot make more of them. People can use many renewable energy resources. People have used energy from the sun, wind, moving water, biomass, and Earth for thousands of years. Modern technology gives us new ways to use these renewable resources.

Essential Vocabulary

- biomass energy   page 39
- geothermal energy   page 38
- hydroelectric power   page 36

▼ Many homes are designed to take advantage of solar energy.

What are the advantages and disadvantages of renewable energy resources?

CHAPTER 4

Solar Energy

Energy from the sun was one of the first energy resources people used. Ancient people built thick-walled homes. The walls absorbed sunlight during the day and then released that heat into the home at night. People also used sunlight to heat water and dry food.

A lot of sunlight reaches Earth's surface. Twenty days of solar energy have as much energy as all of the fossil fuel reserves on Earth.

CHECKPOINT

Make Connections

Think about how the sun influences your life. Could life go on without the sun?

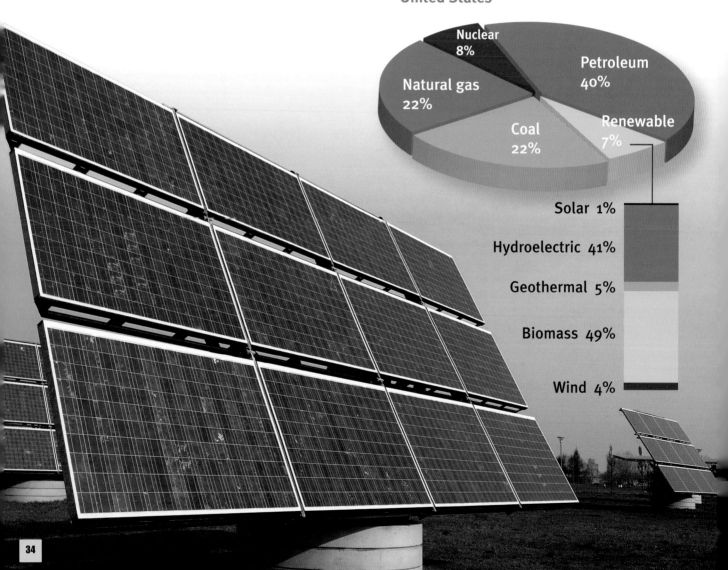

▼ Percentage of energy produced by each energy source in the United States

- Nuclear 8%
- Petroleum 40%
- Natural gas 22%
- Coal 22%
- Renewable 7%

Solar 1%
Hydroelectric 41%
Geothermal 5%
Biomass 49%
Wind 4%

Have you ever seen solar panels on a roof? Solar panels have solar cells that collect sunlight and make electricity. Electricity from solar energy is called solar power. So far, solar cells are expensive and turn only a small part of sunlight into electricity. We must store solar power in batteries when the weather is not sunny.

That is why solar power costs more today than electricity from burning fossil fuels. But people are improving solar technology to make it cheaper. Solar panels already cost less than power lines in places that are far away.

Advantages and Disadvantages of Solar Energy

ENERGY RESOURCE	ADVANTAGES	DISADVANTAGES
Solar Energy	• Produces no air pollution or carbon dioxide • Enormous supply • Renewable	• Only available on sunny days • Solar cells inefficient and expensive • Amount of sunlight is not constant at a given location • Requires large surface areas to collect the energy to be used at a useful rate

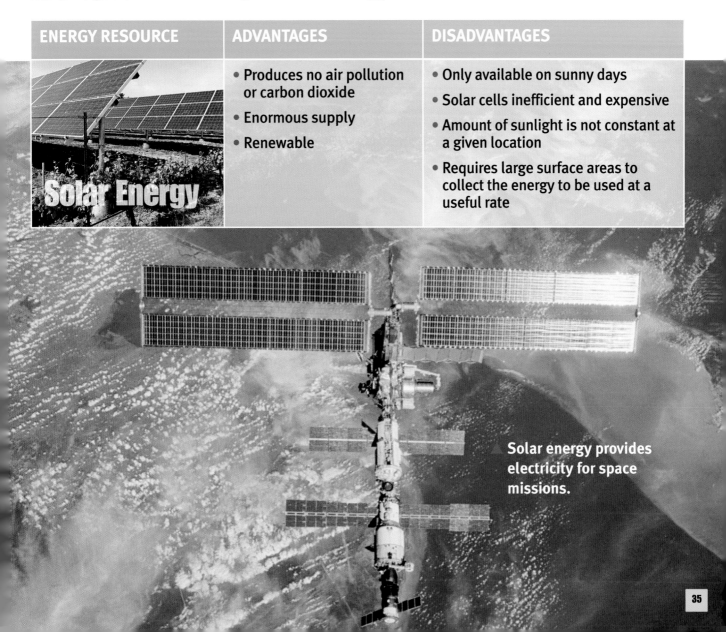

Solar energy provides electricity for space missions.

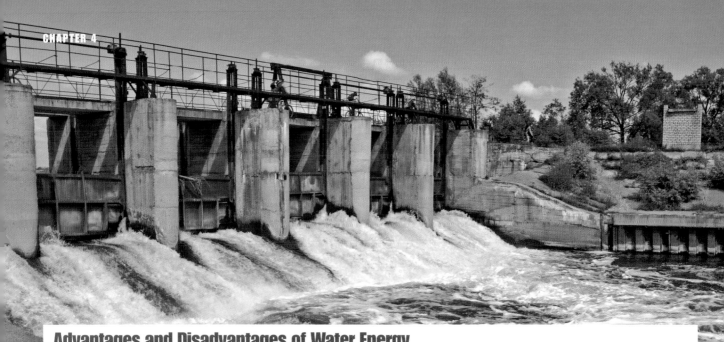

Advantages and Disadvantages of Water Energy

ENERGY RESOURCE	ADVANTAGES	DISADVANTAGES
Water Energy 	• Produces no air pollution or carbon dioxide after dam is complete • Relatively inexpensive electricity • Enormous supply • Renewable	• Limited sites for dams • Environmental impact of dams and flooding • Building dams releases carbon dioxide • Can disrupt plant and animal habitats • Sometimes can force relocation of people

Water Energy

Flowing water is a type of kinetic energy. People used the energy of moving water to turn wheels and run mills over 2,000 years ago. Now, dams collect water and use it to spin turbines that make electricity, or **hydroelectric power** (HY-droh-ih-LEK-trik POW-er). Moving water costs nothing and makes no pollution or carbon dioxide.

Why don't we get all of our energy from moving water? First, we do not have enough rivers to dam. People are trying to make hydroelectric power plants that use the energy of ocean waves and tides.

Second, dams impact the environment. Making concrete for a dam releases a lot of carbon dioxide. This carbon dioxide causes global climate change. Dams flood the areas behind them. States have begun to destroy some hydroelectric dams because they harm fish.

Wind Energy

Moving air is also a type of kinetic energy. The wind's kinetic energy spins the turbines. This makes electricity. Large wind farms make enough electricity to power whole towns. Farming can also continue below the turbines. Wind power produces no pollution. Wind turbines cost less to build than dams or power plants.

What are the limitations of wind energy? Some people do not like the way wind turbines look. They can kill birds and bats flying by. The greatest problem is that people do not know when the wind will blow. This means people cannot always depend on wind. Wind power works best when combined with other energy sources.

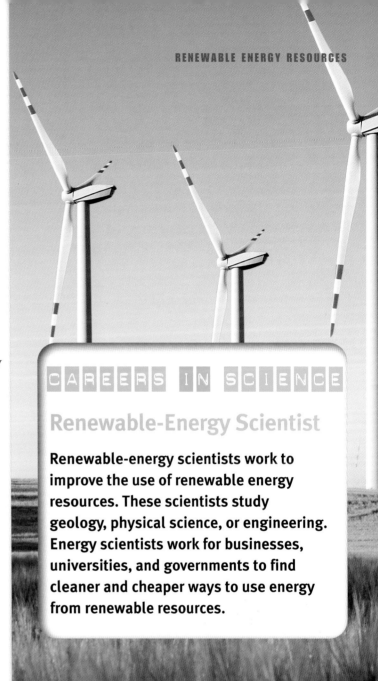

CAREERS IN SCIENCE

Renewable-Energy Scientist

Renewable-energy scientists work to improve the use of renewable energy resources. These scientists study geology, physical science, or engineering. Energy scientists work for businesses, universities, and governments to find cleaner and cheaper ways to use energy from renewable resources.

Advantages and Disadvantages of Wind Energy

ENERGY RESOURCE	ADVANTAGES	DISADVANTAGES
Wind Energy	• Produces no air pollution or carbon dioxide • Enormous supply • Renewable • Wind turbines can share land with other uses	• Wind not always steady or predictable • Wind turbines can harm wildlife • Some people find wind turbines unattractive

CHAPTER 4

Geothermal Energy

It is very hot inside Earth. This **geothermal energy** (jee-oh-THER-mul EH-ner-jee) comes from the molten rock below Earth's surface. It also comes from decaying radioactive elements. In some places, the hot rocks are close to Earth's surface.

Power plants use geothermal energy to boil water and make electricity. First, people pump water underground. Heat energy from the hot rocks turns the water into steam. Steam turns a turbine to make electricity. Gases released from Earth are the only pollution from geothermal power.

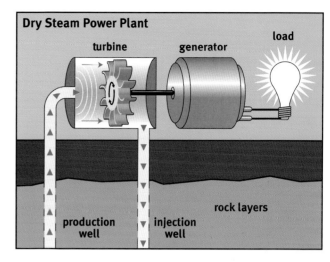

We can get geothermal energy only in certain places. Iceland makes 27% of its electricity from geothermal energy. Iceland also uses geothermal energy for most of its heating. Geothermal power plants in California, Hawaii, Nevada, and Utah give electricity to millions of homes in the United States.

Advantages and Disadvantages of Geothermal Energy

ENERGY RESOURCE	ADVANTAGES	DISADVANTAGE
Geothermal Energy	• Releases very little pollution or carbon dioxide • Enormous supply • Renewable	• Only available where hot rocks lie close to Earth's surface

Geothermal power plants use Earth's heat to generate electricity.

Biomass Energy

Biomass energy (BY-oh-mas EH-ner-jee) is the potential energy in plant material and animal waste. Biomass is a renewable resource because its energy comes from the sun. Burning wood, plant stalks, and animal wastes makes biomass energy. This energy heats buildings and makes electricity. People also turn biomass into biofuels.

Plants use carbon dioxide from the air to grow. Burning biomass or biomass fuel releases that carbon dioxide back into the air. The use of biomass energy produces no new carbon dioxide.

Biomass fuels are often made from food crops such as corn, soybeans, and sugarcane. Biofuel could reduce the amount of food for people and livestock. Other sources of biomass are inexpensive or free. It is difficult to turn grass or cornstalks into biofuels.

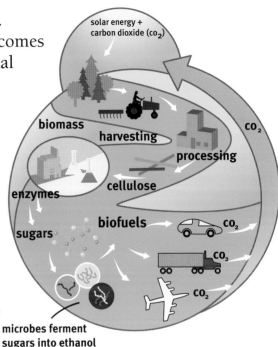

The Biomass Fuel Cycle

Advantages and Disadvantages of Biomass Energy

ENERGY RESOURCE	ADVANTAGES	DISADVANTAGES
Biomass Energy	• Fuels made from waste products are inexpensive • Produces no new carbon dioxide	• Most plant material is difficult to convert to fuel • Competition for land between food and fuel uses

SCIENCE AND TECHNOLOGY
Genetic Engineering

Plants contain tough fibers. Grass-eating animals such as cows use bacteria to break down plant fibers into sugars. Scientists are using genetic engineering to make these bacteria useful to people. The engineered bacteria could digest grass and wood into sugars for biomass fuel production.

CHAPTER 4

They Made a Difference
Amory Lovins (1947 –)

Amory Lovins is an American physicist who studies energy resources. He believes that the wise use of energy resources is very important. Lovins argues that business and the environment both benefit when people reduce energy use. His ultimate goal is sustainable energy use—energy use that does not have a negative impact on the environment or the people of the future. It will take renewable energy resources to achieve sustainable energy use.

Lovins founded the Rocky Mountain Institute (RMI) to help companies and governments make sensible energy choices. RMI helped the Empire State Building in New York City reduce its energy use by nearly 40%. Lovins is also working to develop a vehicle that gets 100 miles per gallon of gasoline. Lovins is famous for finding solutions that reduce energy use and increase profits at the same time.

▶ Amory Lovins works with businesses to reduce their energy needs.

"Our energy future is choice, not fate." — Amory Lovins

SUMMING UP

RENEWABLE ENERGY RESOURCES

- The sun, wind, water, biomass, and heat of Earth are all renewable energy resources.
- Nature quickly replaces these energy resources.
- Energy produced from renewable resources has the advantages of producing little pollution and little or no carbon dioxide.
- However, each renewable resource has limitations such as cost or limited availability.
- Renewable energy resources are a small but growing part of the world energy picture.

Putting It All Together

Choose one of the activities below.

1. Imagine that your state passed a law requiring 20% of its electricity to come from renewable resources. Which renewable resources make the most sense in your state? Create a list of resources.

2. Write a paragraph responding to the following statement: "Renewable energy resources are the only energy resources with no environmental costs." Trade papers with a classmate and discuss your ideas.

3. Make a poster showing all five types of renewable energy in use. Share your poster with your class.

CONCLUSION
Quest for Energy

The Three Gorges Dam is now the largest hydroelectric dam in the world.

Buildings in the future may use multiple energy resources along with energy-saving features.

The Chinese began building the Three Gorges Project in 1994. In 2003, the hydroelectric power plant began to make electricity. The dam now makes more electricity than a dozen nuclear power plants and replaces the burning of forty to fifty million tons of coal each year. Selling the electricity will pay for the dam by 2017.

The Three Gorges Dam also caused many problems. Making the concrete for the dam gave off a lot of carbon dioxide. The rising waters behind the dam made millions of people move. Changes in the Yangtze River made the Chinese river dolphin extinct.

How will we power our lives in the future? We will need a mix of several energy resources. People are thinking of ways to use every energy resource in cleaner, cheaper, and safer ways. Not using so much energy is the cleanest and safest way to use energy resources.

How to Write a Persuasive Letter

Have you ever felt so passionate about a subject that you wanted to let everyone in your community know about it? You can spread the word with a letter to the editor of your school or local newspaper. The newspaper may print your persuasive letter on the editorial page or the newspaper's Web site.

Step 1 Choose a topic that is important to you and to your community. For example, you might want solar-powered lighting installed in a new skate park.

Step 2 Carefully read the guidelines for letters to the editor. You can find these guidelines on the newspaper's Web site or editorial page.

Step 3 Collect facts and figures that you want to present. Make an outline of your most important points.

Step 4 Write your letter as clearly and briefly as possible. State your main point in the first sentence. Follow your main point with a sentence or two giving background information. End by suggesting what the reader can do to help.

Step 5 Type your letter and proofread it carefully. Keep your tone respectful. A well-written, polite letter will convince more readers than an angry, careless note.

Step 6 Sign your name and give any required contact information. Submit your letter by mail or e-mail.

Sample Letter

To the Editor:

Deerwood's new skate park is missing one important detail: lighting. I believe that the park would be much safer with lights around the edges. Without lighting, accidents will be more likely between dusk and the park closing at 8 P.M.

Solar-powered lighting is ideal for the skate park. The lights are only needed for a couple of hours each day. Solar lights do not require any power lines or extension cords. During the day, the lights store energy from sunlight. In the evening, the lights come on automatically and shine for a few hours. Solar lampposts cost about the same as traditional streetlights. After purchase, solar lights cost nothing to operate.

Deerwood residents can contact their council member about adding lights to the skate park. Together, we can save money and keep skaters safe.

Sincerely yours,
Freddie Marzano

Glossary

biomass energy (BY-oh-mas EH-ner-jee) *noun* potential energy found in plant material and animal waste (page 39)

energy (EH-ner-jee) *noun* the ability to do work (page 7)

energy resource (EH-ner-jee REE-sors) *noun* an energy source used to meet the needs of people (page 8)

fission (FIH-shun) *noun* the splitting of one atomic nucleus into two nuclei with the release of a lot of energy (page 27)

fossil fuel (FAH-sul FYOOL) *noun* an energy resource that is formed inside Earth from ancient plant or animal remains (page 16)

geothermal energy (jee-oh-THER-mul EH-ner-jee) *noun* energy from inside Earth (page 38)

hydroelectric power (HY-droh-ih-LEK-trik POW-er) *noun* electricity generated from the energy of moving water (page 36)

kinetic energy (kih-NEH-tik EH-ner-jee) *noun* energy of movement (page 7)

nonrenewable resource (nahn-rih-NOO-uh-bul REE-sors) *noun* a natural resource that cannot be replaced (page 8)

nuclear power (NOO-klee-er POW-er) *noun* electricity generated using nuclear energy (page 28)

nucleus (NOO-klee-us) *noun* the core of an atom containing protons and neutrons (page 26)

potential energy (puh-TEN-shul EH-ner-jee) *noun* stored energy (page 7)

radioactive (ray-dee-oh-AK-tiv) *adjective* spontaneously releasing energy and particles from an atomic nucleus (page 27)

renewable resource (rih-NOO-uh-bul REE-sors) *noun* a resource quickly replaced by nature (page 8)

Index

biomass energy, 39

chain reaction, 27

coal, 4, 8, 12, 16, 19–20, 23, 28, 42

efficiency, 13

energy, 4, 6–8, 11–13, 15, 18, 24–25, 27–28, 31, 33–34, 36–37, 39–41, 43, 45

energy resource, 4, 8, 10–13, 15–16, 19–23, 30, 33–34, 40–41, 43

fission, 27, 31

fossil fuels, 8, 10, 13, 16, 18, 22–23, 30, 33, 35

geothermal energy, 38

global climate change, 13, 19, 22, 36

hydroelectric power, 36, 43

kinetic energy, 7–8, 11, 28, 36–37

Lovins, Amory, 40

natural gas, 8, 16, 18, 23

nonrenewable resources, 8, 16, 22–23, 28

nuclear energy, 8, 27–28, 30–31, 33

nuclear meltdown, 30

nuclear power, 28, 30–31, 43

nuclear waste, 30–31

nucleus, 26–28

petroleum, 4, 8, 10, 12, 15–16, 18, 21–23

pollution, 4, 13, 19, 22, 28, 31, 36–38, 41

potential energy, 7–8, 11, 39

radioactive, 27–28, 30, 38

renewable resources, 8, 33, 37, 39, 41

solar energy, 8, 28, 34–35

solar power, 35

Three Gorges Project, 4–5, 43

water energy, 36

wind energy, 37